Fire Apparatus/Train Collision
Catlett, Virginia

Department of Homeland Security
United States Fire Administration
National Fire Data Center

U.S. Fire Administration Fire Investigations Program

The U.S. Fire Administration develops reports on selected major fires throughout the country. The fires usually involve multiple deaths or a large loss of property. But the primary criterion for deciding to do a report is whether it will result in significant "lessons learned." In some cases these lessons bring to light new knowledge about fire--the effect of building construction or contents, human behavior in fire, etc. In other cases, the lessons are not new but are serious enough to highlight once again, with yet another fire tragedy report. In some cases, special reports are developed to discuss events, drills, or new technologies which are of interest to the fire service.

The reports are sent to fire magazines and are distributed at National and Regional fire meetings. The International Association of Fire Chiefs assists the USFA in disseminating the findings throughout the fire service. On a continuing basis the reports are available on request from the USFA; announcements of their availability are published widely in fire journals and newsletters.

This body of work provides detailed information on the nature of the fire problem for policymakers who must decide on allocations of resources between fire and other pressing problems, and within the fire service to improve codes and code enforcement, training, public fire education, building technology, and other related areas.

The Fire Administration, which has no regulatory authority, sends an experienced fire investigator into a community after a major incident only after having conferred with the local fire authorities to insure that the assistance and presence of the USFA would be supportive and would in no way interfere with any review of the incident they are themselves conducting. The intent is not to arrive during the event or even immediately after, but rather after the dust settles, so that a complete and objective review of all the important aspects of the incident can be made. Local authorities review the USFA's report while it is in draft. The USFA investigator or team is available to local authorities should they wish to request technical assistance for their own investigation.

This report and its recommendations were developed by USFA staff and by TriData Corporation, Arlington, Virginia, its staff and consultants, who are under contract to assist the USFA in carrying out the Fire Reports Program.

The USFA greatly appreciates the cooperation received from the National Transportation Safety Board (NTSB), the Catlett Volunteer Fire Company, the Fauquier County Office of Emergency Services, the Fauquier Fire/Rescue Association, and the National Railroad Passenger Corporation.

For additional copies of this report write to the U.S. Fire Administration, 16825 South Seton Avenue, Emmitsburg, Maryland 21727. The report is available on the USFA Web site at http://www.usfa.dhs.gov/

U.S. Fire Administration

Mission Statement

As an entity of the Department of Homeland Security, the mission of the USFA is to reduce life and economic losses due to fire and related emergencies, through leadership, advocacy, coordination, and support. We serve the Nation independently, in coordination with other Federal agencies, and in partnership with fire protection and emergency service communities. With a commitment to excellence, we provide public education, training, technology, and data initiatives.

TABLE OF CONTENTS

Fire Apparatus/Train Collision
Catlett, Virginia
September 1989

"The United States Fire Administration dedicates this report to the many firefighters and emergency medical personnel who responded to this tragic train collision and derailment. The report is published in memory of Firefighter Jay Mark Miller and Firefighter Matthew Brian Smith of the Catlett Volunteer Fire Department who lost their lives in the line of duty."

— Olin L. Greene, United States Fire Administrator

INTRODUCTION

This report is a summary of the investigation reports prepared by the NTSB about this incident. The NTSB reports are based on NTSB's investigation, as well as information provided by the Catlett Volunteer Fire Company, the Fauquier County Office of Emergency Services, Fauquier Fire/Rescue Association, and the National Railroad Passenger Corporation. Staff of the United States Fire Administration (USFA) Office of Firefighter Health and Safety assisted the NTSB with the investigation of the collision. This is Report 048 of USFA's Major Fires Investigation Project and was compiled by Charles Jennings of TriData Corporation under contract EMW-90-C-3338.

The USFA has produced this report in accordance with the recommendations of the NTSB following their investigation. Specifically, the NTSB recommended that the USFA:

- Notify fire companies of the facts and circumstances of the Fire Apparatus/Train Collision that occurred near Catlett, Virginia, on September 28, 1989, and urge those companies to develop, implement, and periodically review and practice plans to safely cross railroad grade crossings during an emergency response. Any plans should emphasize that the safe arrival of the apparatus at the scene of the emergency is the first priority.

The NTSB also made a recommendation to the National Fire Protection Association (NFPA) to ensure that this issue is appropriately addressed in National consensus standards, and to Operation Lifesaver to ensure that the Nation's emergency service vehicles are included in railroad grade crossing public information and education programs.

Shortly after the release of the NTSB investigation reports, the NTSB released "Special Investigation Report: Emergency Fire Apparatus." For this report, the NTSB examined eight separate fire apparatus accidents and conducted an informal survey of the 50 States and the District of Columbia to determine their requirements for inspecting fire apparatus. The safety issues discussed in the report are fire department vehicle maintenance programs and State inspection programs, fire department operating procedures concerning manual brake limiting valves and engine retarders, and fire apparatus occupant seatbelt use. Recommendations concerning these issues were made to the USFA, the International Association of Fire Chiefs (IAFC), the NFPA, and those States which do not have existing programs in-place to periodically inspect fire apparatus.

1

Specifically, the following recommendation were made

...to the USFA of the Federal Emergency Management Agency:

- Urge fire departments to establish vehicle maintenance programs that follow all of the manufacturer's service requirements and schedules.

- Inform fire departments nationwide of the potential hazards of misusing engine retarders and encourage fire departments to establish operating procedures that are consistent with manufacturer's warnings about the proper use of engine retarders.

- Notify fire departments of the hazards of using fire apparatus manual brake limiting valves and urge them to discontinue the use of these devices.

- In cooperation with NFPA and IAFC, encourage fire departments to establish and enforce mandatory seatbelt policies and to develop programs that promote the use of seatbelts in fire apparatus.

...to the IAFC:

- Urge fire departments to establish vehicle maintenance programs that follow all of the manufacturer's service requirements and schedules.

- Inform fire departments nationwide of the potential hazards of misusing engine retarders and encourage fire departments to establish operating procedures that are consistent with manufacturer's warnings about the proper use of engine retarders.

- Notify fire departments of the hazards of using fire apparatus manual brake limiting valves and urge them to discontinue the use of these devices.

- Cooperate with the USFA and the NFPA to encourage fire departments to establish and enforce mandatory seatbelt policies and to develop programs that promote the use of seatbelts in fire apparatus.

...to the NFPA:

- Cooperate with the USFA and the IAFC to encourage fire departments to establish and enforce mandatory seatbelt policies and to develop programs that promote the use of seatbelts in fire apparatus.

...to the governors and legislative bodies of those States without fire apparatus inspection programs:

- Develop and implement a fire-apparatus inspection program that requires periodic inspections performed by commercial vehicle inspectors in accordance with the Federal Highway Administration Motor Carrier Assistance Program vehicle (mechanical) inspection criteria.

This USFA Special Investigation Report details the circumstances of and emergency response to the fire apparatus/train collision that occurred near Catlett, Virginia, on September 28, 1989. It is recommended that fire departments obtain the NTSB Special Investigation Report on Emergency Fire Apparatus for more information on the circumstances leading to the above recommendations.

OVERVIEW

On September 29, 1989, an engine, Wagon 7, of the Catlett (Virginia) Volunteer Fire Department was struck by an Amtrak passenger train while crossing a private grade crossing in a rural area. The engine was responding to a vehicle fire and was in sight of smoke from the burning vehicle when the accident occurred. Two of the five firefighters on the engine were killed, the other three were injured – two seriously. Of the 399 passengers and crew on the Amtrak train, 57 were injured, with minor to moderate injuries, including the entire train crew. The accident completely destroyed the pumper and caused the two locomotives and first 11 cars of the train to derail leaving five cars on the track. Property damage to the train, railroad right-of-way, and the pumper were estimated at over 1 million dollars.

The collision ruptured the gasoline tank on Wagon 7 and caused a gasoline fire that burned under the first several cars of the train. These fires were extinguished by other firefighters who were responding to the vehicle fire and witnessed the collision.

The emergency response involved fire apparatus from six counties and included over 15 engines, 38 ambulances, eight rescue squads, seven helicopters, and other equipment from fire departments in six counties, six police departments, the U.S. Army, and the Washington Hospital Center.

The Catlett Volunteer Fire Department had a practice in-place whereby operators of fire apparatus were instructed to stop their vehicles before crossing an unprotected railroad crossing. The National standards on driver training and professional qualifications in effect at the time of the collision did not address the issue of rail crossings, and this topic is not often covered in related courses. The NTSB suggested that standards for rail crossings be developed and implemented by National fire service organizations, and that fire departments develop and implement plans to safely cross railroad grade crossings.

As part of the effort to publicize this information, the USFA has released this report on the incident. According to USFA data, there were seven grade crossing accidents nationwide which resulted in the deaths of nine firefighters from 1977 through 1988.

THE COLLISION

At 1920 hours on September 28, 1989, the Fauquier County Sheriff's Department dispatched the Catlett Volunteer Fire Department for a reported vehicle fire about one mile south of Catlett on State Route 28. Shortly after Wagon 7 checked on the air at 1923 hours, the chief of the Catlett Volunteer Fire Department ascertained that the engine had left the station without an officer on board, contrary to department policy. The chief radioed the responding unit, Wagon 7, and asked them if they had a full crew on board. Instead of answering the question, they requested that any additional units be held until they arrived at the scene. The chief agreed, although he recognized from the voice on the radio that there was no officer on the unit.

At 1935 hours, Wagon 7 requested that the Fauquier County Emergency Operations Center repeat directions to the fire. The directions were given again. Hearing this request, Chief 7 determined that Wagon 7 had missed the entrance to the fire site. The chief and a lieutenant, who was at the station, checked on the air and responded to the car fire. The chief responded in his private vehicle and the lieutenant responded in a tanker (Tanker 7) with two other firefighters. After their response was announced, Wagon 7 radioed the chief to report that they missed the driveway. They stated "We passed it, we're coming back from Calverton (a community about 1.6 miles south of the fire.)" The chief responded, "I hope so."

The chief arrived at the scene of the fire at 1937 hours. A diagram of the accident scene is included in Appendix A. After driving up a 59 foot driveway and crossing a single set of railroad tracks, the chief drove about one quarter mile further up the driveway and reported that the vehicle was "fully involved." Tanker 7 had left the station about one minute after the chief. As they approached the scene, they observed Wagon 7 coming from the opposite direction. Its emergency equipment was operating and both the driver and other firefighter in the cab were looking in the direction of the fire. At the entrance to the driveway, the tanker stopped to allow Wagon 7 to enter the driveway first. Wagon 7 overshot the driveway and had to back up to align their vehicle to make the turn into the driveway. The lieutenant and the other firefighters in Tanker 7 reported that they did not see brake lights come on as Wagon 7 drove up the driveway toward the fire. The lieutenant in Tanker 7 heard two horn blasts, which he first thought to be coming from another apparatus responding to the fire. The lieutenant and firefighter in Tanker 7 observed the headlight of an approaching train. Tanker 7 observed the train strike Wagon 7 as it was attempting to cross the tracks.

The impact destroyed Wagon 7 and derailed much of the train. The gasoline tank on Wagon 7 ruptured and the gasoline ignited, forming a fireball and flame that traveled down the tracks as the train pushed the wreckage of Wagon 7 approximately 970 feet past the crossing. The driveway from Route 28 to the vehicle fire was blocked by derailed train cars. All but the last five cars of the 18-car train derailed. Several passenger cars came to rest in a tilted position as the train telescoped after the collision. None of the cars overturned, however.

THE TRAIN

Amtrak train number 19, "The Crescent," en route from Washington, D.C. to New Orleans, Louisiana, had collided with Wagon 7. There were 379 passengers and 20 crew members aboard the train. The train consisted of two locomotives, a baggage/crew dormitory car, two baggage cars, and 13 passenger cars.

The event recorder aboard the locomotive indicated that the train was traveling about 77 miles per hour before the collision occurred. The engineer stated that he observed the fire apparatus and believed that Wagon 7 was going to stop at the crossing. When the engine entered the crossing, the engineer applied the brakes on "emergency" and sounded the horn. The engineer reported that the firefighter riding in the front passenger seat of Wagon 7 never looked at the train before the collision.

The brunt of the collision was directed at the rear of the vehicle, at about the rear axle. When the collision occurred, the cab and chassis of the Catlett pumper rotated counterclockwise 450 degrees and came to rest about 80 feet southeast of the crossing. The lead locomotive stopped about 965 feet past the crossing with the left side of Wagon 7's hose body wrapped around its front end. Most equipment and the rear bodywork were scattered along the collision area. Gasoline fires that broke out near the second locomotive and several derailed cars were extinguished by other firefighters responding to the emergency. The primary fuel for these fires was from the fuel tank aboard Wagon 7, which ruptured in the crash.

FACTORS CONTRIBUTING TO THE COLLISION

The Crossing

The crossing where the collision occurred was not equipped with any automatic warning equipment such as flashing lights or gates. The only warning was a standard cross buck (railroad crossing) sign. The driveway led from Route 28 to a private residence.

Even though it was dusk, the weather conditions at the time of the accident were clear. The 59-foot section of the driveway from the highway to the rail crossing had an 11.9 degree grade. But the tracks were straight and there were no visual obstructions that prevented the driver of Wagon 7 from seeing the approaching train. NTSB investigators estimate that the railroad tracks are visible for approximately 3,700 feet in the direction the train came from.

Driver Stress

The driver of Wagon 7 was a 24-year-old member of the department who had several years of experience in operating large vehicles. He had undergone training by the fire company in opera-tion of the vehicle and had been driving Wagon 7 for about three years. Postmortem interviews and toxicological tests revealed no evidence of any physical impairment.

In addition to normal stressors experienced during emergency response, the actions of the driver of Wagon 7 indicated he was undoubtedly experiencing added stress for several reasons. First, the engine left the station without an officer on board, contrary to department policy. The chief's radio transmissions asking "who is in charge" and the indirect response from the crew of Wagon 7 would indicate that the driver was under added stress from the initial moments of the response. Wagon 7 radioed the chief to request that any additional apparatus be held in the station until they arrived at the scene to verify the nature of the call.

A second contributing factor, which may have been affected by the first, was the fact that Wagon 7 missed the turn for the driveway leading to the fire and traveled approximately 1.5 miles past the fire before asking for directions from Fauquier County communications. When Wagon 7 requested directions to the fire, the chief and Tanker 7 responded to the scene. Despite leaving the station four minutes after Wagon 7, the chief arrived on the scene before any other apparatus and reported that the vehicle was fully involved. Upon hearing this, it is likely that the crew of Wagon 7 focused their attention on "redeeming themselves"[1] by performing satisfactorily in reaching and extinguishing the fire. The tanker and the engine approached the scene at the same time. Tanker 7 stopped to allow Wagon 7 to enter the driveway first. The heightened level of stress on the driver of Wagon 7 is indicated by the fact that as he approached the driveway leading to the location of the fire call, he overshot the turn and had to back the vehicle to make the turn into the driveway. In recreations of the maneuver with a similar apparatus, it took 16 seconds to properly align the apparatus to move up the driveway.

At this point, the car fire was visible to the driver of Wagon 7 and the crew probably focused all their attention on reaching the fire. This is reinforced by the statements of the Amtrak engineer, who stated that the placed the train's brakes on emergency and sounded the horn when he realized that the vehicle was not going to stop. As the train headed toward the pumper, the engineer stated that

---

[1] NTSB memo (H90-112) to FEMA Director, page 4.

the front seat passenger never looked at the train, although the passengers in the rear jump seats did observe the approaching train.

"Although some level of stress can enhance human performance, excessive stress can lead to substandard performance. When a person's arousal level is unduly increased by stressors, the focus of attention is narrowed to performance of the task perceived to be the most important, while the quality of the performance of any peripheral task(s) deteriorates."[2]

Interestingly, the Catlett Volunteer Fire Department had responded to a car struck by a train at a similar crossing several months before this incident, which would indicate that they should have been aware of the dangers associated with such crossings.

FINDINGS RELATED TO RAIL CROSSING PROCEDURES

Wagon 7 was a 1978 Ford/Oren pumper equipped with a 1000--GPM pump, 750-gallon tank, and Ford V-8 gasoline engine. The apparatus was believed to be in good repair at the time of the incident. The NTSB investigators recreated the events preceding the collision with the assistance of a similar Ford/Oren pumper purchased in 1979 by the Remington Volunteer Fire Department. Among the tests performed were weighing the vehicle, determining the engine noise at various speeds, and reenacting the view of the railroad tracks from the cab of the vehicle.

Among the findings of the NTSB investigation of this accident were:

- If it is not practical to plan an emergency response route that avoids grade crossings, selection of crossings that are equipped with automatic warning devices is preferable to selection of those that are not. All planning should include identification of the location at the crossing from which a driver or other observer assigned to the apparatus can see the maximum available distance down the track(s) on both sides.

- Train horns may not be audible when a vehicle siren is operating. Engine and cab noise may be sufficient to obscure the sound of train horns.[3] It may be necessary for drivers of large vehicles to stop the vehicle; idle the engine; turn off all radios, fans, wipers, and other noise-producing devices in the cab; lower the window; and listen for a train's horn before entering the grade crossing.

- A "challenge-response" protocol should be used when negotiating an unprotected rail crossing. After stopping the vehicle, the driver and passenger should ascertain that there is no train coming from either direction before proceeding. The challenge and response method would positively assure that the driver was taking proper precautions during the response.

- Because trains are not required to sound their horns at private grade crossings, it is necessary to visually determine that no train is approaching before entering a crossing. Reliance cannot be placed on listening for the train horn.

[2]Wickens, C.D., "Engineering Psychology and Human Performance," Charles E. Merrill Publishing Co., University of Illinois at Champaign-Urbana, 1984, 20156-X, pp. 249-290. As cited in memo H-90-112.

[3]NTSB Safety Study: "passenger/Commuter Train and Motor Vehicle Collisions at Grade Crossings (1985), " NTSB/SS-86/04, 1986. Originally cited in NTSB memo (H-90-112) to FEMA Director.

- In cases where an approaching train could be obscured from direct observation from the safely positioned vehicle, such as crossings located on curves in the railroad track or multiple track crossings where a stopped train blocks vision, it may be necessary to have a member of the crew proceed ahead of the apparatus on foot to assure that there is no train approaching before signaling the vehicle to cross the tracks.

The window on the driver's side of Wagon 7 was open at the time of the crash. Investigators were not sure if the siren on Wagon 7 was operating, but the tanker, which was immediately behind Wagon 7 did have its siren operating. Tests performed on a similar apparatus indicated that the engine noise from driving up the grade to the crossing would also have been loud enough to interfere with hearing. The attention of both the driver and officer were focused on the car fire they were responding to and not on assuring that the railroad crossing was clear before they drove across the tracks.

This incident also demonstrates the need for fire departments to plan their response routes to minimize the number of unprotected rail crossings they must pass to reach an emergency. The NTSB findings can be used as guidelines for establishing policies for operation over grade crossings within their response areas.

POST-CRASH INCIDENT MANAGEMENT

With the Chief of Company 7 and the tanker crew on either side of the accident, an immediate call was placed to Fauquier County fire communications to notify them of the collision. Firefighters from Tanker 7 moved their apparatus to extinguish fires that erupted near the second locomotive. These firefighters initially used several fire extinguishers and eventually, a two and a half inch line to extinguish the fires under the derailed cars. The fire involving the remains of Wagon 7's gasoline tank took considerably longer to extinguish. The prompt extinguishment of these fires was critical in limiting casualties among the train crew and passengers.

An ambulance responding for the car fire arrived immediately following the collision and began searching for the crew of Wagon 7 while firefighters from Tanker 7 extinguished the fires. The firefighters from Wagon 7 had been ejected from the vehicle and were located on the far side of the railroad tracks from Route 28, and the rescuers had to crawl under and between the derailed cars to reach them. The rescuers determined that two of the firefighters were dead and three were in varying states of consciousness. The two most seriously injured firefighters were transported by helicopter to trauma centers in Fairfax County, Virginia, and Washington, D.C. The third firefighter was transported by ambulance to Fauquier County Hospital, where he was treated and released.

Chief 7 notified Fauquier County Communications Center of the collision between his engine and the train and immediately began to request additional units. Within seven minutes of the collision, an ambulance that responded to the scene requested that helicopters in the area be put on standby for evacuation of crash victims. At ten minutes into the Incident a Command Post was established and the National Fire Academy Incident Command System (NFA ICS) was implemented.

Among the staff on the units responding to the request for help was the Fauquier County Emergency Services Director. When he arrived on the scene, he relieved the Catlett Volunteer Fire Department Chief of command and established a Command Post on Route 28 opposite the derailed passenger cars. He immediately notified Fauquier County to alert area hospitals of the situation.

Rescue and emergency medical services (EMS) crews began systematically removing people from the train cars after stabilizing those cars which had derailed. Crews used railroad ties displaced by the incident to crib underneath those cars that were unstable. Those passengers who had escaped on their own were guided to a field across Route 28 from the accident site to ensure their safety and accountability. Injured patients were also removed to the same area for secondary triage and treatment. Disposition and transportation to appropriate medical facilities were coordinated from this area, as was the transportation of uninjured passengers by school bus to a local school that had been opened as an emergency shelter.

With the incident occurring along a two-lane section of highway in a rural area, the response of emergency vehicles to the incident quickly created traffic problems. Route 28 was the main north-south artery and was loaded with normal traffic in addition to emergency responders. The added influx of onlookers, the media, and pedestrian traffic adjacent to the collision created severe traffic problems. Local and State police were eventually able to alleviate this problem.

Law enforcement and Amtrak officials worked through the night with local hospitals and EMS personnel to account for the status and disposition of all passengers and crew members. This was an arduous task, as many victims were treated and released at area hospitals before an accountability system was established; others left the hospital before being treated. Eventually, all 399 people originally aboard the train were accounted for.

Despite the difficult operating conditions and the magnitude of the operation, there were no injuries to fire service or EMS personnel responding to this incident following the collision. One school bus driver who was transporting uninjured passengers from the train suffered chest pains and was transported to a hospital.

COMMUNICATION PROBLEMS

Dispatch Center

The scope of this incident quickly overwhelmed the capabilities of the Fauquier County Sheriff Department's Communications Center. This center dispatches for county fire and rescue departments, as well as the sheriff's department. Staffed by only two personnel, they were faced with dispatching units, relaying information, notifying other jurisdictions and hospitals, requesting assistance, and handling inquiries from the media and public – all in addition to handling other emergencies throughout the county.

These dispatchers were eventually supplemented by three dispatchers sent by neighboring Loudoun County. Fairfax County sent a mobile command unit equipped with a phone, radio, computer, and logistics capabilities which responded to the scene and relieved the sheriff's department of some duties.

Radio Frequency Coordination

Because of the number of jurisdictions and agencies working together on this incident, there were problems in establishing communication with incoming and arriving units. The incident involved fire and rescue departments from six counties and two military installations; police from three jurisdictions, including Amtrak; helicopters from two police departments, one hospital, and the US. Army; and ten school buses from Fauquier County. Among the problems experienced was that no direct communication was possible with the U.S. Army helicopters, for example, which prevented them from being safely used.

Public Information

The initial information dissemination effort at the site was handled by a Fauquier County Sheriff's deputy. However, a designated press/public information area was not established. Some information given out to the media was unconfirmed and was later found to be erroneous. A public information officer (PIO) from the Virginia State Police arrived and took over the duties of PIO, and was later assisted by members of the Fairfax County Fire and Rescue Department Public Affairs section. After some initial confusion, the three agencies developed a strong information network to keep the media and the public informed. The State police retained control over public information duties for several days following the incident.

Fauquier County Disaster Plan

Fauquier County was in the process of revising its disaster plan, which was originally written in 1983. While significant progress had been made, the plan had not been officially disseminated to the fire and rescue departments in the county. The presence of the County Emergency Services Director as Incident Commander enabled several new procedures to be implemented.

One key success in the incident was the adoption of the NFA's ICS by fire and rescue departments in the county. Because this system is in use throughout the Region, mutual aid units were successfully integrated into the incident without undue confusion. A more rapid implementation of ICS would have helped organize the initial stages of the incident.

CRITICAL INCIDENT STRESS DEBRIEFING (CISD)

A part of the draft disaster plan was the use of CISD teams. Recognizing the potential for emotional trauma in this incident which killed two firefighters and injured three others, and injured 77 civilians, CISD teams were requested early in the incident. Fire and rescue personnel from Catlett and Cedar Run (the other company on the initial response to the car fire) were relieved of duty as soon as practical, and were sent to meet with trained peers and psychologists from the Prince William County CISD team to minimize any long-term emotional problems.

Other response personnel were debriefed by a team from State Planning District 9 in Culpeper, Virginia. The passengers aboard the train were debriefed by a team of experts from Fauquier County. With serious incidents, this counseling can continue for many weeks, but it is most effective when initiated soon after a traumatic incident occurs. As a result of planning, members of the initial response crew received this assistance as quickly as possible.

In addition, the damaged pumper was covered up by another fire company and taken to their station to relieve Catlett firefighters of the emotional pain associated with seeing it.

LESSONS LEARNED

The Collision

1. **Driver training programs should include instruction on proper procedures for crossing railroad crossings, especially those without automatic warning devices.**

   This critical behavior is not included in National fire service driving or professional qualifications standards such as "NFPA 1002," Fire Apparatus Driver/Operator Professional Qualifications.[4] As a result, many training classes designed to meet the objectives of this standard do not include information on safely negotiating rail crossings when driving fire apparatus.

   The Virginia State standard that applies to fire apparatus driving is in the form of a curriculum entitled "Virginia Department of Fire Programs Standard Training Curricula: Virginia Emergency Vehicle Operators Course." This training program is designed to meet the requirements of NFPA 1002 and therefore has the same deficiencies. Although this program is taught throughout the State, it is not a mandatory program for fire apparatus drivers.

2. **Fire departments should consider the importance of having an officer on board apparatus whenever possible.**

   While many departments do not require that an officer is aboard every piece of apparatus, this incident indicates that in this case, the absence of an officer may have contributed to the collision between the fire department vehicle and the train.

3. **The stress of emergency response can impair driving ability.**

   The NTSB investigation revealed that the events leading up to the collision between Wagon 7 and the Amtrak train added stressors to the task of vehicle operation. The crew left the station without an officer, was discovered by the chief, missed the turn for the fire, and was subsequently beaten to the fire by their chief, who left the station several minutes after they did. These factors contributed to the driver's and front seat passenger's failure to stop their vehicle and look for an approaching train before they entered the crossing.

The Emergency Response

1. **Good safety procedures can keep a bad situation from getting worse.**

   In this incident, victim safety and rescuer safety were good. Despite the stressful nature of this incident and the difficult operating conditions, there were no injuries to fire/rescue personnel responding to the train wreck. Given the number of units involved, this indicates that good safety procedures were followed by those personnel responding. Rapid shoring of partially-overturned passenger cars allowed for safe egress of occupants, and the evacuation of the train's passengers was accomplished without further injury to them.

---

[4] NFPA 1002 is designed to build on the requirements of NFPA 1001, *Firefighter Professional Qualifications*. Other NFPA standards of relevance to apparatus driving are NFPA 1021, *Fire Officer Professional Qualifications*, NFPA 1500, *Fire Department Occupational Safety and Health Programs*, and NFPA 1901, *Automotive Fire Apparatus*. While all of these standards relate to driver safety, none of the editions in effect at the time of the collision contained any reference to procedures for negotiating unprotected rail crossings.

2. **Routine use of ICS, planning, and multi-agency exercises are needed for adequate conditions of command staff.**

The initial difficulties in organizing an incident command structure for an incident of this magnitude could be reduced through routine use of ICS at all incidents and through planning and the use of multi-agency disaster exercises. Given that the county disaster plan was being revised at the time of the incident, the operation went well once an ICS was implemented.

3. **There is a need for emergency management personnel to respond to the Emergency Operations Center in incidents of this magnitude.**

In this incident, the Fauquier County Emergency Services Director responded directly to the incident. While this was helpful for some aspects of utilizing the newly-developed portions of the county disaster plan, there was a need for emergency management personnel to be at the communications center to provide assistance and coordination of resources, that is, to be at a fixed site away from the incident.

4. **Need to set up press area and PIO immediately at incident of this magnitude.**

Incident Commanders should anticipate the media and public interest that will arise in response to an incident such as this. The lack of a designated PIO and an established press area at the Catlett incident led to erroneous media reports. The impact of media inquiries on dispatch centers should also be considered when planning for these incidents.

5. **Plan in advance for communications between agencies in disaster situations.**

Aside from the fact that the dispatch center was overwhelmed by duties related to this incident, there were several responding agencies which had no communications capability with the Incident Commander. These problems should be identified in the planning stage and arrangements made to provide communications or utilize these resources in a capacity where direct communication is not necessary.

BIBLIOGRAPHY

The following sources were used in preparation of this report:

National Transportation Safety Board, *Safety Recommendation* (H-90-112), addressed to FEMA Director, January 4, 1991.

National Transportation Safety Board, *Safety Recommendation* (H-90-113), addressed to NFPA Standards Council, January 4, 1991.

National Transportation Safety Board, *Safety Recommendation* (H-90-114), addressed to Operation Lifesaver, January 4, 1991.

National Transportation Safety Board, *Special Investigation Report: Emergency Fire Apparatus* (SIR-91/01) March 19, 1991.

National Transportation Safety Board, Bureau of Technology, *Survival Factors Specialist's Factual Report*, NTSB no. DCA-89-MH-001.

APPENDIX A

Diagram Showing Final Rest Positions of Wagon 7 and Railroad Cars, and
Chart of Type of Rail Car, Number, and Order in Train

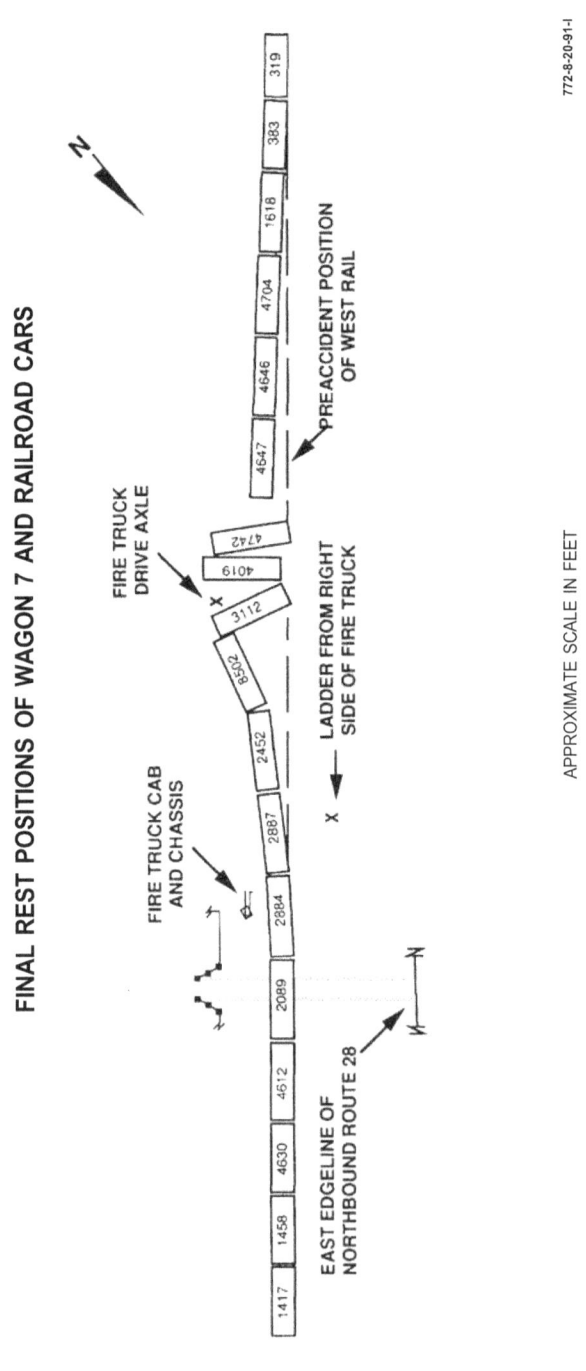

FINAL REST POSITIONS OF WAGON 7 AND RAILROAD CARS

12

Appendix A (continued)

Rail Car Positions and Numbers

Order in Train	Type of Car	Exterior Car Number
1	Locomotive	319
2	Locomotive	383
3	Baggage Car	1618
4	Coach	4704
5	Coach	4646
6	Coach	4647
7	Coach	742
8	Coach	4019
9	Lounge	112
10	Diner	8502
11	Sleeper	2452
12	Sleeper	2887
13	Sleeper	2884
14	Sleeper	2089
15	Coach	4612
16	Coach	4630
17	Mail Car	1458
18	Mail Car	1417

APPENDIX B

Photographs

Photo supplied by Mr. Clyde Lomax

Left side of Wagon 7

14

Appendix B (continued)

Wagon 2 at accident crossing – camera facing north

Appendix B (continued)

NTSB photo

Front of locomotive 319

Appendix B (continued)

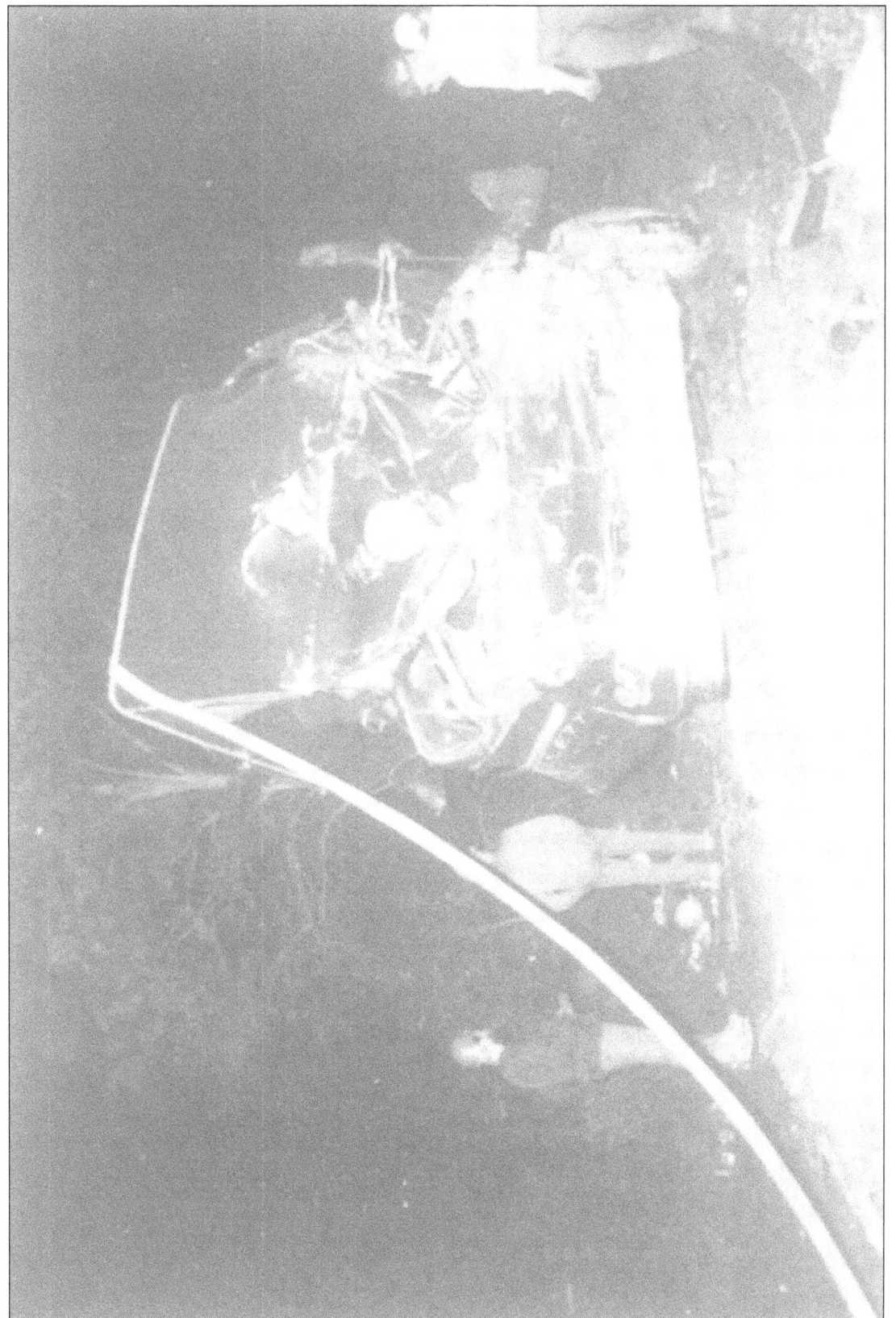

NTSB photo

Cab of Wagon 7 – camera facing southeast

Appendix B (continued)

NTSB photo

Cab of Wagon 7

Appendix B (continued)

NTSB photo

Cab and chassis of Wagon 7
Southwest corner of fence around private residence is in the center foreground.

NOTE: Injured firefighters were found in the area at the bottom left of the photo.

Appendix B (continued)

NTSB photo

Chassis and cab of Wagon 7

Appendix B (continued)

Amtrak photo

Closeup of fire truck drive axle

Appendix B (continued)

NTSB photo

Right front of locomotive 319

Appendix B (continued)

Amtrak photo

West ends of cars 3112, 4019, and 4742

Appendix B (continued)

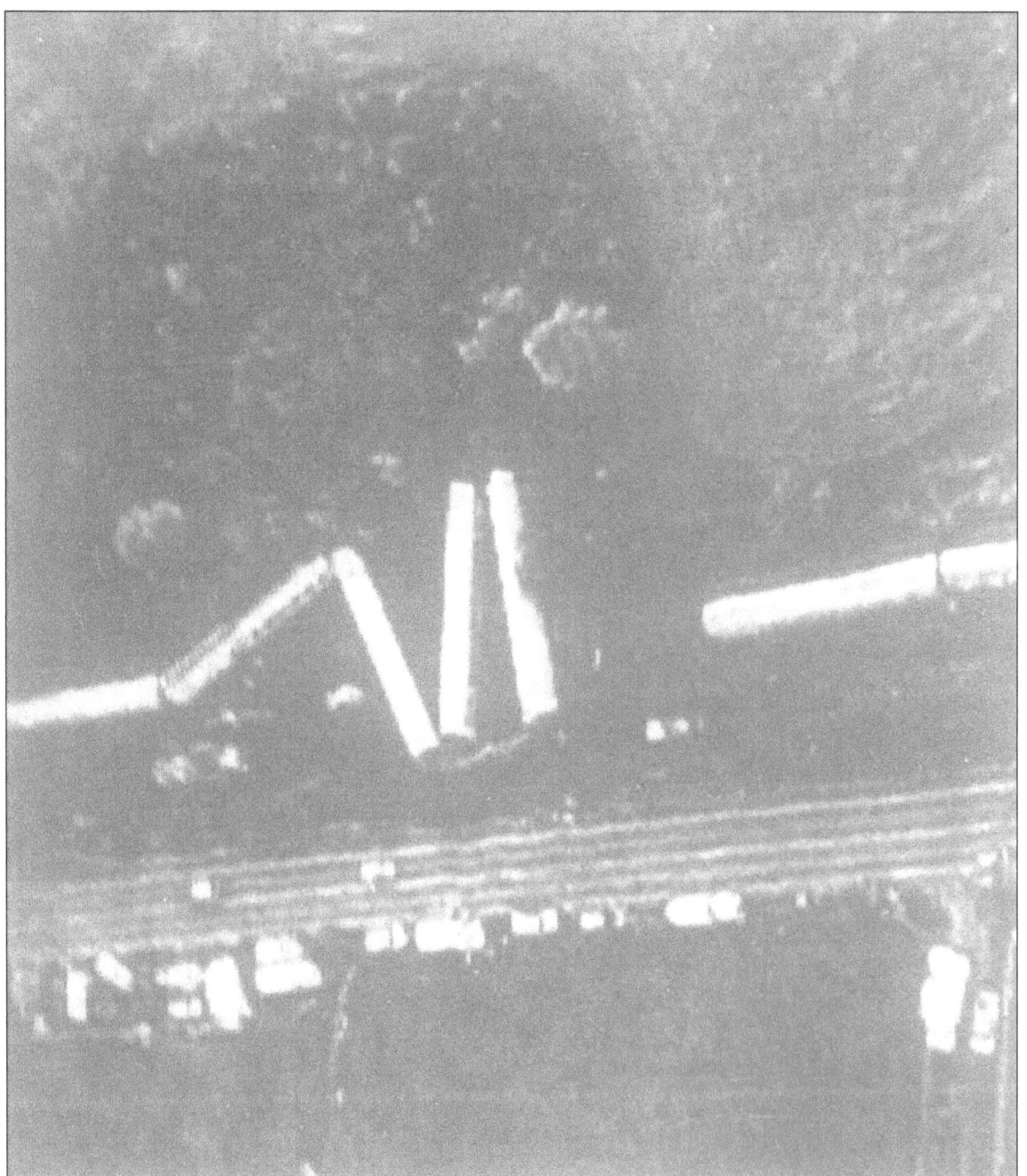

Photo from WTTG-TV videotape

Aerial view of accident site

Appendix B (continued)

NTSB photo

North side of car 4742

Appendix B (continued)

Fauquier OES photo

South side of car 4742

Appendix B (continued)

Fauquier OES photo

Evacuation operations from west end of car 4742

Appendix B (continued)

Fauquier OES photo

West end of car 4742

Appendix B (continued)

Fauquier OES photo

Evacuation of passengers from west ends of cars 4019 and 3112

APPENDIX C

Summary of Response:
Departments, Equipment, Command/Staff Personnel

Fire and Rescue Departments

Fauquier County
Prince William County
Culpeper County
Loudoun County
Fairfax County
Stafford County
Fairfax City

Police Departments

Virginia State Police
Fauquier County Sheriff
Prince William County Police
Amtrak Police

Medivac Helicopters

Fairfax County Police ("Fairfax 2")
United States Park Police ("Eagle 1")
Washington Hospital Center (3 "Medstar" helicopters)
United States Army (2 helicopters from Fort Belvoir)

Fire and Rescue Units (by type)

15 pumpers
4 tankers
1 ladder truck
8 rescue squads
3 SERV units*
24 ambulances
14 medic units
2 light units
2 cave-in units

1 mobile command unit
7 medivac helicopters
10 buses for evacuation
3 critical incident stress debriefing teams
39 command/staff personnel:
      18 chief officers
      10 EMS supervisors
      6 training officers
      3 dispatchers
      2 public information officers

Fauquier County –
Initial dispatch time: 19:40

6 pumpers
2 tankers
1 rescue squad
2 SERV units*
3 ambulances
8 medic units
10 staff personnel:
      5 chief officers
      5 EMS supervisors

Prince William County –
Initial dispatch time: 19:46

7 pumpers
2 tankers
4 rescue squads
1 ladder truck
1 light unit
10 ambulances
2 medic units

*SERV unit (Special Emergency Response Vehicle): a van-type vehicle which has communication and support capabilities, and is used for other specialized support-type functions as determined by the various departments.

Appendix C (continued)

14 staff personnel:
> 10 chief officers
> 2 EMS supervisors
> 2 training officers

Culpeper County –
Initial dispatch time: 19:52

1 pumper
3 ambulances
1 EMS supervisor

Fairfax County –
Initial dispatch time: 20:03

1 pumper
3 rescue squads
5 ambulances
2 medic units
2 canteen units
1 light unit
1 mobile command unit
9 staff personnel:
> 2 chief officers
> 1 EMS supervisor
> 4 training officers
> 2 public information officers

Loudoun County –
Initial dispatch time (approximate):
20:30

4 staff personnel:
> 1 chief officer
> 3 dispatchers

Stafford County –
Initial dispatch time: 20:41

1 SERV unit
2 ambulances
1 medic unit
1 EMS supervisor

Fairfax City –
Initial dispatch time (by Fairfax
County): 21:20

1 ambulance
1 medic unit

APPENDIX D

Radio Log

Fauquier (Lavoie):

19:27:54 (Two tones.) Engine Company 7, a vehicle fire, Route 28 just south of Catlett, 19:27. (Two tones.) Engine Company 7, a vehicle fire, Route 28 just south of Catlett. 19:28. KBD 680.

Company 7 (Smith):

19:32:00 Company 7.

Fauquier (Lavoie):

19:32:03 Engine Company 7. A vehicle fire, reported to be well involved, Route 28 at the Farthing Farm approximately a mile south of Catlett on the left-hand side.

Company 7 (Smith):

19:32:13 Ten-four. Wagon 7's responding.

Fauquier (Lavoie):

19:32:16 Wagon 7 responding. 19:32.

19:32:19 Wagon 7 (Smith):

Fauquier, give me an off tone please.

Fauquier:

19:32:29 (Two tones.)

Chief 7 (Lomax):

19:32:43 Chief, 7, Wagon 7.

(No response to Chief 7 recorded.)

Company 7 (Anderson):

19:32:49 Company 7 to Wagon 7.

Wagon 7 (Smith):

19:32:52 Wagon 7.

Company 7 (Anderson):

19:32:54 You all want the tanker?

Wagon 7 (Smith):

Appendix D (continued)

19:32:56   Yeah. Go ahead and bring it.

Chief 7 (Lomax):

19:33:06   Chief 7, Wagon 7.

Wagon 7 (Smith):

19:33:08   Wagon 7.

Chief & (Lomax):

19:33:09   You have a full crew?

Wagon 7 (Smith):

19:33:11   Ten-four.

Chief 7 (Lomax):

19:33:14   Are you in charge?

(No response to Chief 7 recorded.)

Wagon 7 (Miller):

19:33:36   Wagon 7 to Company 7.

Company 7 (Lomax):

19:33:43   Go ahead.

Wagon 7 (Miller):

19:33:46   You can hold off on the tanker 'til we get on the scene. We'll let you know what we got.

Company 7 (Lomax):

19:33:50   Okay.

Wagon 7 (Smith):

19:35:00   Wagon 7 to Fauquier.

Fauquier (Lavoie):

19:35:02   Wagon 7.

Wagon 7 (Smith):

19:35:03   Have a set of directions, please?

Fauquier (Lavoie):

19:35:07   The best the caller could advise me was the Farthing Farm, approximately a mile south of Catlett on the left. I believe it's going to be the first residence on your left past the Cedar Run Bridge. Be the Steel residence.

Appendix D (continued)

Wagon 7 (Smith):

19:35:19   All right. Ten-four. We're direct.

Fauquier (Lavoie):

19:35:23   19:35

Tanker 7 (Shrock):

19:36:17   Tanker 7's responding, Fauquier.

Fauquier (Lavoie):

19:36:20   Okay, Tanker 7. 19:36.

Chief 7 (Lomax):

19:36:23   Chief 7 also.

Fauquier (Lavoie):

19:36:25   Chief 7 responding. 19:36.

Chief 7 (Lomax):

19:36:29   Tanker, uh, Chief 7, Tanker 7, first driveway across the Cedar Run Bridge on the left, that new house.

Tanker 7 (Shrock):

19:36:36   I'm direct, Chief

Wagon 7 (Smith):

19:36:40   Wagon 7 to Chief 7.

Chief 7 (Lomax):

19:36:42   Go ahead, Wagon.

Wagon 7 (Smith):

19:36:43   We passed it. We're coming back from Calverton.

Chief 7 (Lomax):

19:36:44   I hope so.

Chief & (Lomax):

19:37:03   Chief 7 to Tanker 7. When you get up the new house come straight on back. It's fully involved.

Tanker 7 (Shrock):

19:37:10   Okay, Chief.

Appendix D (continued)

Chief 7 (Lomax):

19:37:33   Chief 7 on location, Fauquier. Vehicle's fully involved.

Fauquier (Lavoie):

19:37:38   Chief 7 on the scene. Vehicle fully involved.

19:37

COLLISION OCCURS (STATIC BURST AT 19:37:55)

Tanker 7 (Shrock):

19:38:11   Tanker 7 to Chief 7. Are you aware of what just happened?

Chief 7 (Lomax):

19:38:14   What happened? Didn't hit my truck, did it?

Tanker 7 (Shrock):

19:38:19   That's right. We have heavy fire here along the tracks.

Shock-Trauma 12-6 (McDevitt):

19:38:26   Shock-Trauma 12-6 is on the scene, Fauquier. Give me a couple more ambulances.

Chief 7 (Lomax):

19:38:31   Chief 7, Fauquier, be advised one of my trucks has been hit by a train. Give me a full response.

Tanker 7 (Shrock):

19:38:41   Lieutenant 7 to Chief 7, we're going to need more than that. We have a passenger train involved here.

Chief 7 (Lomax):

19:38:47   Chief 7 Fauquier.

Fauquier (Lavoie):

19:38:49   Chief 7.

Chief 7 (Lomax):

19:38:51   Give me a 3 alarm. I've got a passenger train involved, a car involved, and my fire truck involved. Give me several ambulances and a rescue squad ASAP.

Fauquier (Lavoie):

19:38:59   19:38

Shock-Trauma 12-6 (McDevitt):

Appendix D (continued)

19:39:02   Shock-Trauma 12-6 Company 12.

(No response to Shock-Trauma 12-6 recorded.)

Chief 2 (Mason):

19:39:07   This is Rescue Chief 2. Go ahead and roll the two units in my station, please.

Chief 7 (Lomax):

19:39:12   Chief 7 to Lieutenant 7.

(No response to Chief 7 recorded.)

Fauquier (Lavoie):

19:39:16   (Two tones.) Rescue and Engine Company 2, Rescue Company 6, Engine Company 7, Engine Company 13, and Rescue Company 12, fire apparatus struck by a train, Route 28, in Catlett, 19:39. (Eleven tones.) Rescue and Engine Company 2, Rescue Company 6, Engine Company 7, Rescue Company 12, Engine Company 13, fire apparatus struck by a train, Route 28 in Catlett. Squad 6 also due. 19:40 KBD 680.

www.ingramcontent.com/pod-product-compliance
Lightning Source LLC
Chambersburg PA
CBHW081237170526
45165CB00009B/3084